Cars, Trucks, and Bikes

Cars, Trucks, and Bikes

By Steve Parker

Illustrated by Alex Pang

MASON CREST PUBLISHERS INC.
370 Reed Road, Broomall, Pennsylvania 19008
(866)MCP-BOOK (toll free), www.masoncrest.com

First Printing
9 8 7 6 5 4 3 2 1

Library of Congress Cataloging-in-Publication Data
Parker, Steve, 1952–
Cars, trucks, and bikes / by Steve Parker ; illustrated by Alex Pang.
    p. cm. — (How it works)
Includes bibliographical references and index.
ISBN 978-1-4222-1792-4
Series ISBN (10 titles): 978-1-4222-1790-0
1. Automobiles—Juvenile literature. 2. Trucks—Juvenile
literature. 3. Motorcycles--Juvenile literature. 4. Bicycles—
Juvenile literature. I. Pang, Alex, ill. II. Title.
TL147.P373 2011
629.2—dc22
            2010033614

Printed in the U.S.A.

First published by Miles Kelly Publishing Ltd
Bardfield Centre, Great Bardfield, Essex, CM7 4SL
© 2009 Miles Kelly Publishing Ltd

Editorial Director: *Belinda Gallagher*
Art Director: *Jo Brewer*
Design Concept: *Simon Lee*
Volume Design: *Rocket Design*
Cover Designer: *Simon Lee*
Indexer: *Gill Lee*
Americanizer: *Paula Lewis*
Production Manager: *Elizabeth Brunwin*
Reprographics: *Stephan Davis, Ian Paulyn*
Consultants: *John and Sue Becklake*

Every effort has been made to acknowledge the source and
copyright holder of each picture. The publisher apologizes for any
unintentional

ACKNOWLEDGMENTS

All panel artworks by Rocket Design
The publishers would like to thank the following
sources for the use of their photographs:
Alamy: 17 DBURKE; 29 Motoring Picture Library
Corbis: 7 (b) Clifford White; 11 Mike King;
21 Diego Azubella/epa; 23 Chris Williams/Icon/SMI;
25 Transtock; 26 Walter G. Arce/ASP Inc.Icon SMI;
35 Thinkstock; 36 George Hall
Fotolia: 15 Sculpies – Fotolia.com
Getty Images: 30 Tim Graham; 33 James Balog
Rex Features: 7 (c) The Travel Library; 9;
13 KPA/Zama; 19 Motor Audi Car
Science Photo Library: 6 (t) LIBRARY OF CONGRESS;
All other photographs are from Miles Kelly Archives

WWW.FACTSFORPROJECTS.COM

Each top right-hand page directs
you to the Internet to help you
find out more. You can log on
to **www.factsforprojects.com**
to find free pictures, additional
information, videos, fun
activities, and further web links.
These are for your own personal
use and should not be copied or
distributed for any commercial
or profit-related purpose.

If you do decide to use the
Internet with your book, here's a
list of what you'll need:
• A PC with Microsoft® Windows®
  XP or later versions, or a
  Macintosh with OS X or later,
  and 512Mb RAM

• A browser such as Microsoft®
  Internet Explorer 8, Firefox 3.X,
  or Safari 4.X
• Connection to the Internet via
  a modem (preferably 56Kbps) or
  a faster Broadband connection
• An account with an Internet
  Service Provider (ISP)
• A sound card for listening to
  sound files

Links won't work?
www.factsforprojects.com is
regularly checked to make sure
the links provide you with lots
of information. Sometimes you
may receive a message saying
that a site is unavailable. If this
happens, just try again later.

Stay safe!
When using the Internet, make
sure you follow these guidelines:
• Ask a parent or guardian's
  permission before you log on.
• Never give out your personal
  details, such as your name,
  address, or e-mail.
• If a site asks you to log in or
  register by typing your name or
  e-mail address, speak to your
  parent or guardian first.
• If you do receive an e-mail from
  someone you don't know, tell
  an adult and do not reply to the
  message.
• Never arrange to meet anyone
  you have talked to on the
  Internet.

CONTENTS

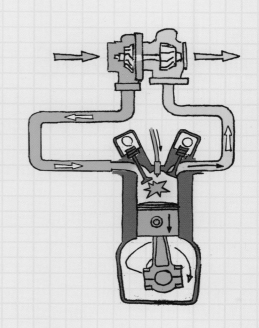

INTRODUCTION. 6

MOUNTAIN BIKE. 8

ROAD RACE BICYCLE 10

TOURING MOTORCYCLE 12

SUPERBIKE 14

SEDAN CAR. 16

SUPER SPORTS CAR. 18

F1 RACING CAR 20

DRAGSTER 22

4WD OFF-ROADER 24

RALLY CAR 26

PICKUP TRUCK. 28

CITY BUS. 30

SEMITRUCK 32

TOW TRUCK 34

FIRE TRUCK 36

GLOSSARY. 38

INDEX 40

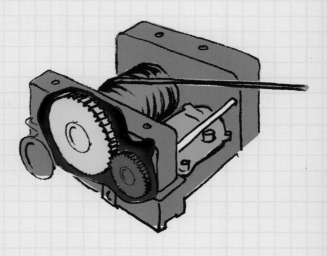

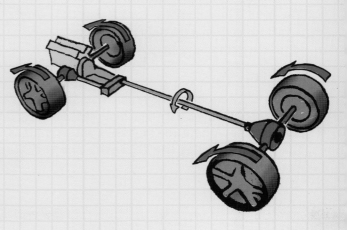

INTRODUCTION

The wheel was invented about 6,000 years ago in western Asia. Initially, it was a potter's wheel, used for shaping clay bowls and vases. By 3,500 years ago, wheels were in use on wagons and chariots pulled by horses, oxen, or slaves. It took another 3,300 years to invent wheeled vehicles with engines. Prior to this, the first bicycles appeared. They had no pedals. Riders pushed their feet against the ground. Personal engine-driven transport came next. We have never looked back—except to see who is behind us.

The penny-farthing bicycle of the 1870s had direct drive with pedals attached to the wheel.

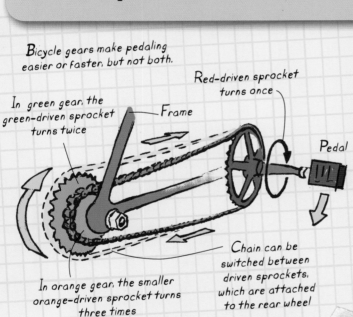

Bicycle gears make pedaling easier or faster, but not both.

In green gear, the green-driven sprocket turns twice

Red-driven sprocket turns once

Frame

Pedal

In orange gear, the smaller orange-driven sprocket turns three times

Chain can be switched between driven sprockets, which are attached to the rear wheel

ON THE ROAD

The first pedal-powered bicycles and engine-driven cars and motorcycles appeared toward the end of the 1800s. Most transport was still animal-drawn and roads were little more than dirt tracks with sharp stones and deep holes. Many early cars were steam- or electric-powered, and most were hand-built in the tradition of horse-drawn carriages.

MOTORING FOR ALL

In 1908, the American Ford Motor Company introduced an assembly line where many identical cars could be put together from already-made parts. Suddenly, vehicles were cheaper and demand grew. By the 1950s, some cars were huge, covered with shiny chrome, and lined with leather. Real mass motoring took hold in the 1960s with smaller budget cars such as the VW Beetle and the Austin Mini.

The British Mini was not only a very small car, but it was also a fashion item and symbol of the 1960s.

The vehicles featured in this book are Internet linked.
Visit www.factsforprojects.com to find out more.

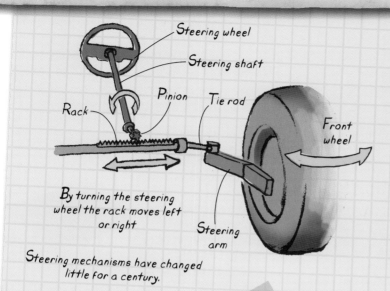

Steering wheel

Steering shaft

Pinion

Rack

Tie rod

Front wheel

By turning the steering wheel the rack moves left or right

Steering arm

Steering mechanisms have changed little for a century.

BRANCHING OUT

As road vehicles became more powerful and more reliable, they evolved into new kinds. Today, there are huge trucks to transport loads, emergency vehicles such as fire trucks, service trucks, all-wheel-drive off-roaders, and pickup trucks. More gears, better brakes, stronger engines, and smoother suspensions help to make these vehicles more efficient and the ride more comfortable. Whether driving the family car, riding about on bicycles, boarding the school bus, or cruising the empty freeway, vehicles are vital in our daily lives.

Massive road trains are used to transport goods across the vast open areas of Australia.

THE RACE IS ON!

Any new way of going places meant that people wanted to be there first. Cycling races and motor sports blossomed, from rallying to track events to dragster duels. In today's Formula One races, super-fast cars that cost tens of millions of dollars race to the biggest audiences on the planet.

Motor vehicles have come a long way in a century, but with fossil fuels running out and global warming on the increase, could we live without them?

More than 600 million TV viewers watch each Formula One race.

MOUNTAIN BIKE

The overall design of the bicycle has hardly changed for more than 100 years. It is made up of many simple machines or basic mechanical devices such as levers, wheels and axles, pulleys, gears, and springs. A bicycle also gives its rider exercise to stay healthy, and because it has no engine and polluting exhaust gases, it's excellent for the environment.

Mountain biking became an official Olympic sport in 1996—100 years after ordinary cycle racing.

Eureka!

The first bicycles were developed in Germany and France during the 1810s. However, they had no pedals. Riders had to push the ground with their feet to scoot along.

What next?

Electric bicycles, complete with a see-through, bubble-like cover to keep the rider dry, are predicted to become more popular.

Spring suspension When the rear wheel goes over a bump, its frame tilts up and squeezes a large spring to absorb the shock.

Gear changer A device called the derailleur moves the chain sideways from one sprocket to another to change gear.

Medium-sized green driven sprocket turns two times (lower gear)

Large red driver sprocket turns once

Frame

Pedal

Driven sprockets are attached to the rear wheel

Small orange driven sprocket turns three times (higher gear)

Chain can be switched between driven sprockets

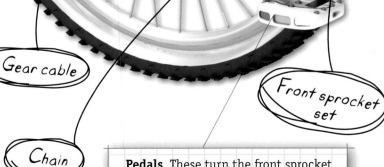

Rear sprocket set

Gear cable

Chain

Front sprocket set

Pedals These turn the front sprocket whose teeth fit into the chain link gaps. This provides a nonslip way to turn the rear sprocket.

✳ How do CHAIN AND SPROCKET GEARS work?

A bicycle's rear sprockets or cogs have different numbers of teeth. In a low gear, each turn of the front sprocket (attached to the pedals) means the rear sprocket (attached to the wheel) turns twice. You don't go very far for one turn of the front sprocket but pedaling is less effort. In a higher gear, the chain moves to a smaller sprocket. This turns three times for one turn of the front sprocket. So you go farther for each turn of the pedals but pedaling is more effort. The idea is to change gear to keep your pedaling speed and effort constant at the best rate for you.

Learn how to use bicycle gears by visiting www.factsforprojects.com and clicking the web link.

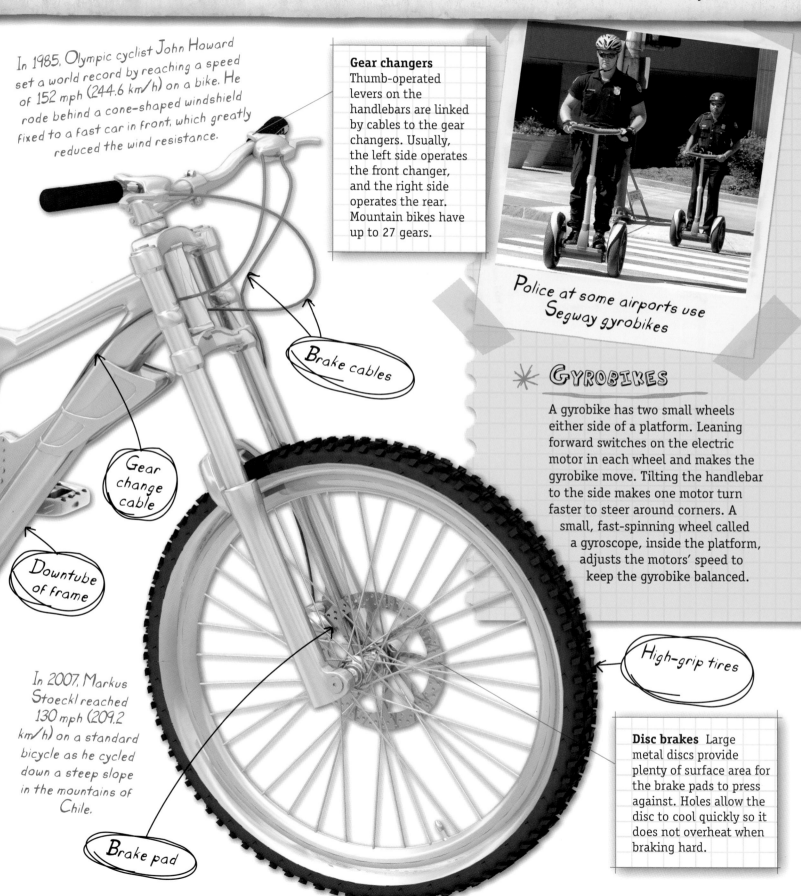

In 1985, Olympic cyclist John Howard set a world record by reaching a speed of 152 mph (244.6 km/h) on a bike. He rode behind a cone-shaped windshield fixed to a fast car in front, which greatly reduced the wind resistance.

Gear changers
Thumb-operated levers on the handlebars are linked by cables to the gear changers. Usually, the left side operates the front changer, and the right side operates the rear. Mountain bikes have up to 27 gears.

Brake cables

Police at some airports use Segway gyrobikes

Gear change cable

✳ GYROBIKES

A gyrobike has two small wheels either side of a platform. Leaning forward switches on the electric motor in each wheel and makes the gyrobike move. Tilting the handlebar to the side makes one motor turn faster to steer around corners. A small, fast-spinning wheel called a gyroscope, inside the platform, adjusts the motors' speed to keep the gyrobike balanced.

Downtube of frame

High-grip tires

In 2007, Markus Stoeckl reached 130 mph (209.2 km/h) on a standard bicycle as he cycled down a steep slope in the mountains of Chile.

Disc brakes Large metal discs provide plenty of surface area for the brake pads to press against. Holes allow the disc to cool quickly so it does not overheat when braking hard.

Brake pad

ROAD RACE BICYCLE

The road race bicycle is the specialist long-distance machine of the cycling world. It's much lighter than a commuter or mountain bike. It is stripped down and simpler, too, with fewer moving parts to go wrong. The best road racers easily cover 120 miles (193 km) in one day, even in bad weather.

Eureka!

The first bicycles had solid wood or metal wheel rims, then solid rubber tires—not very comfortable. The pneumatic or air-filled tire was invented in 1887 by John Boyd Dunlop for his son's bicycle.

Ball bearings A typical bicycle has up to ten sets of ball bearings. Two in the head tube support the handlebars and front forks.

Head tube

Front fork

Racing handlebars Curled-down racing, or dropped handlebars, mean the rider leans forward, head down. This causes less air resistance than sitting upright and allows the legs to press harder on the pedals.

✳ How do BALL BEARINGS work?

A bearing reduces rubbing or friction where one part of a machine moves against another. It decreases wear and lessens the movement energy lost as heat. In a ball bearing design, hard metal balls fit snugly between outer and inner ring-shaped parts called races. The balls can rotate in all directions, which spreads out both the wear over their surfaces and any heat from friction to prevent overheating.

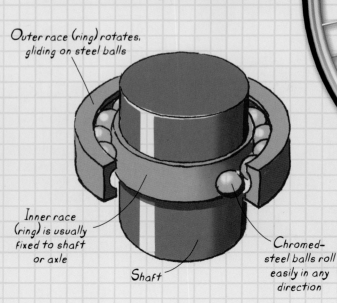

Outer race (ring) rotates, gliding on steel balls

Inner race (ring) is usually fixed to shaft or axle

Shaft

Chromed-steel balls roll easily in any direction

Low-profile tires Narrow and low, the sidewall height is less than the tire width, which gives extra grip but a bumpy ride.

Discover everything you need to know about the world's greatest bicycle race by visiting www.factsforprojects.com and clicking the web link.

Until the 1930s, road-racing bicycles had no gear changers. The rider had to get off and change the rear wheel with a different-sized gear sprocket.

Saddle The long, narrow saddle allows the cyclist's legs to move up and down easily without rubbing on its sides.

In human-powered aircraft, such as *Gossamer Condor* and *Albatross*, pilots use bicycle-like pedals to turn the propellers because the body's strongest muscles are in the leg.

Alloy frame

Down tube

Spokes Most wheels have between 28 and 36 spokes of steel or aluminum alloy. They work by tension or pulling the rim inward, rather than keeping it pushed outward.

The first bicycles with pedals were invented in the 1860s. They turned the front wheel directly without a chain or gears.

Pedal crank

✳ TRACK RACING bikes

Track racing is the Formula One of cycle sport. The bikes are made of the latest composites—mixtures of materials including plastics, metals, and carbon fiber—for the greatest strength with the least weight. Every part of the bike and rider must cause the least air resistance, including the rider's teardrop-shaped helmet.

Wheel rim The rim is made of a lightweight alloy, which is a special combination of several types of metals, including aluminum.

In 1899, Charles Minthorn Murphy was first to pedal one mile in less than a minute—57.75 seconds to be exact.

Track racers keep their heads down to reduce drag

TOURING MOTORCYCLE

Long and low, the touring motorcycle is one of the coolest-looking machines on the road. It may not be the fastest on two wheels, but it cruises the highways in great comfort. The bike has plenty of suspension for a comfortable ride and lots of power for passing and traveling long distances.

Eureka!

In 1885, Gottlieb Daimler fitted his newly designed gasoline engine into the frame of a wooden bicycle and invented the first motorcycle, the Reitwagen (Riding Car). Motorcycles went into mass production in 1894 with the 1500cc Hildebrand & Wolfmüller.

What next?

Electric motorcycles and electric scooters become more popular and faster every year—the speed record is 168 miles an hour (270.4 km/h)!

HARLEY DAVIDSON

Bucket seat

Cylinder cooling fins

Coil spring suspension

Exhaust pipes These long tubes mean exhaust gases are directed out of the engine so that it works more efficiently. They also reduce or muffle the engine's noise.

✳ How does COIL SPRING suspension work?

Spring suspension absorbs the shocks from holes, bumps, and other rough parts of the road. However, when a spring is squeezed and then allowed to push and lengthen again, it tends to bounce back, shortening and lengthening several times. This is reduced by adding a hydraulic damper inside the spring. It's a tube filled with sealed-in oil, into which a smaller tube slides like a telescope. The thick oil slows down and smooths out any fast sliding movements to reduce or lessen the bounce effect.

Upper mounting fixes to motorcycle frame

Oil-filled suspension rod damper reduces the bounce caused by the spring

Strong coil spring compresses as the vehicle goes over a bump

Lower mounting fixes to wheel's suspension arm

Transmission A series of gears transfers the engine's turning power to the drive belt and then on to the rear wheel.

Some motorcycles have a small sidecar for passengers, complete with its own wheel.

Watch a video of a Harley Davidson touring motorcycle by visiting www.factsforprojects.com and clicking the web link.

Throttle A twistgrip on the right handlebar is linked to the engine by a cable. It allows more fuel to enter the engine so it goes faster.

The world-famous Harley Davidson touring motorcycles began in 1903 when friends William Harley and Arthur Davidson produced a one-cylinder version for racing.

Fuel tank The rounded fuel tank in front of the rider is made of very strong metal.

Forks Long front forks allow the front wheel plenty of room to move up when it hits a rough patch of road, which smooths out the ride.

Mudguard

Brake caliper

Brake disc

Engine The two-cylinder 1584cc engine is low down between the two wheels, which makes the motorcycle more stable and less likely to tip over sideways.

✳ What is a MAXI-SCOOTER?

Scooters usually have smaller wheels than motorcycles and streamlined coverings called fairings over most of the vehicle. Older scooters were not very fast or well-balanced. New maxi-scooters are faster, more comfortable, safer, change gear automatically, and can be fitted with a gasoline engine or electric motor.

Maxi-scooters make ideal runabouts

SUPERBIKE

Few road machines can accelerate (pick up speed) or travel as fast as the superbike—a high-powered, souped-up motorcycle. These fierce-looking machines are road versions of even faster, track-racing motorcycles. Because of their great power, light weight, and sensitive steering, they are tricky to ride and definitely not for the beginner.

Eureka!

Before the 1950s, riders had to kick-start their motorcycles by pushing down a pedal to turn over the engine. On most modern bikes, this is done by an electric motor.

What next?

Motorcycle stunt riders are always inventing new tricks—such as speeding off a jump ramp and somersaulting two or three times.

A typical racing superbike can reach an amazing 190 mph (305.8 km/h)—on the track but not on ordinary roads!

Windshield The clear toughened plastic screen makes the air flow up and over the rider at high speed.

Clutch lever The clutch disconnects the engine from the transmission so the rider can change gear without damaging the spinning cogs inside (see page 18).

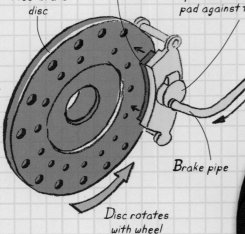

Steel brake disc

Brake pad presses on disc

Hydraulic fluid pushes a piston that presses the brake pad against the disc

Brake pipe

Disc rotates with wheel

✳ How do BRAKES work?

A disc brake has a large metal disc attached to the wheel. Brake pads made of very hard material, such as composite ceramic, press on this to cause friction and slow the disc's rotation. The force to push the pads onto the disc may come from a cable attached to the brake lever. Or it can be hydraulic, from oil that is forced at high pressure from the brake lever or brake pedal along the brake pipe.

Brake disc

Radiator scoop This low opening collects air to flow over the radiator just behind it, which contains the water that keeps the engine cool.

Streamlined fairing

HONDA

CBR

Find out more about how brakes work by visiting
www.factsforprojects.com and clicking the web link.

Since the Superbike World Championships began in 1988, Honda and Ducati have won all but two of the yearly titles.

Apart from being a nickname for a fast bike, Superbike is also an official category of motorcycle racing for engines up to 1200cc.

 QUAD BIKES

A mix of car and motorcycle, quad bikes have four wheels but no main body or covering for the rider. A quad bike is difficult to handle because it tends to tip over if the rider steers around a corner too fast. It's great fun for riding over muddy fields and rough tracks.

Quad bikes have plenty of suspension

Exhaust

Swingarm The U-shaped swingarm tilts up and down at its joint with the frame so the rear wheel can move up and down as part of the suspension.

Tires Racing superbikes have smooth or slick tires with no pattern or tread for good grip at high speed.

Three-spoked wheel

Water-cooled engine

Gear pedal Depending on which gear the motorcycle is already in, flipping the foot pedal up or down changes to the next gear.

Lightweight alloy wheels

SEDAN CAR

The typical sedan car has been around for about 100 years. Its basic design has hardly changed. It has four wheels, an engine in the front that drives the front or rear wheels, two front seats, and seating for two or three in the back seat. It has a separate luggage compartment, or trunk, in the rear.

Eureka!

The first gasoline engine automobile or car (motorized carriage) was built by Karl Benz in 1885. It only had three wheels, the front one was steered by a lever.

What next?

Cars of the future may drive themselves using GPS (satellite navigation) and radio links to a massive central computer.

Engine Typical family cars have engines from 2000cc upward. Usually there are four or six cylinders, one behind the other. The V8 shown here has two rows of four cylinders at a V angle (see page 21).

Front suspension Upper and lower suspension arms allow the front wheels to move up and down.

✳ How do GASOLINE ENGINES work?

A gasoline engine has a piston that moves up and down inside a cylinder. It works in four stages, or strokes.

1. Inlet The piston moves down and draws in a mixture of fuel and air into the cylinder through an inlet valve.

2. Compression The piston moves up and compresses the mixture.

3. Combustion An electric spark plug makes the mixture explode, forcing the piston down.

4. Exhaust The piston goes back up and pushes the burned mixture out through an exhaust valve. The piston's movements turn the engine's crankshaft, by a connecting rod (con-rod).

Headlight

Radiator grill

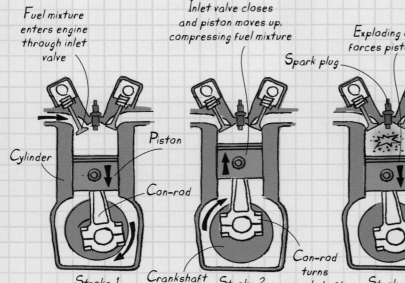

Fuel mixture enters engine through inlet valve

Inlet valve closes and piston moves up, compressing fuel mixture

Exploding mixture forces piston down

Spark plug

Cylinder

Piston

Con-rod

Crankshaft

Con-rod turns crankshaft

Exhaust valve opens, burnt mixture leaves as exhaust gases

Crankcase

Stroke 1

Stroke 2

Stroke 3

Stroke 4

In the early twentieth century, the fastest cars were steam-powered or electric

Watch an animation of the four-stroke engine in action by visiting
www.factsforprojects.com and clicking the web link.

ASTON MARTIN SEDAN

In 1908, the mass-produced Ford Model T or Tin Lizzy meant that ordinary people could afford a car.

Rear transmission The prop shaft is linked by the gears to the rear half-shaft axles on the wheels.

Propeller shaft Many sedans are front-wheel drive. In the rear-wheel drive design (shown here), the propeller shaft carries the turning power from the transmission, along the underside of the car to the rear wheels.

Alloy wheels

The most successful car of all time is the Toyota Corolla, which began production in 1966. More than 35 million Corollas have been sold.

In the 1960s, the tiny Austin Mini was a "must-have" fashion car, with a transverse engine driving the front wheels to save space.

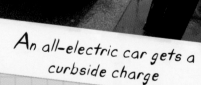

An all-electric car gets a curbside charge

✳ GREEN CARS

Hybrid cars have a small gasoline engine and an electric motor with batteries. The car can run on electricity very quietly with no polluting exhaust fumes. If the batteries run down, the gasoline engine switches on to recharge them. The gasoline engine can also add power to the electric motor for more speed.

SUPER SPORTS CAR

For people with plenty of money, sports cars with powerful engines are the top road machines. They have little cargo space and would scrape against rough terrain, but they are sleek, speedy, and striking. The low design and smooth lines allow the car to move with the least air resistance, which becomes more important as you travel faster.

Eureka!

The first wings appeared on sports cars and racing cars in the 1960s. They work like an upside-down aircraft wing to press the car downward for better tire grip and improved steering.

What next?

Bugatti, maker of the world's leading supercar, the Veyron, plans a new model within five years to hold onto the sports car top spot.

The Bugatti Veyron was introduced in 2005 as the fastest production car in the world. It's also one of the most expensive at approximately $1.5 million.

The Veyron is named after Pierre Veyron, the racing driver who won the 1939 Le Mans 24 Hour race in a Bugatti.

Retractable wing The rear wing is known as a spoiler. To travel at speeds of between 124 and 230 miles an hour (200-370 km/h) in the Veyron, a switch lowers the car and wing, which then disturbs air flow that might pull the car upward.

W16 engine The W16 engine is a double version of the V8 engine. It has 16 cylinders in four rows of four at an angle, like two overlapping Vs.

✳ How does a transmission work?

A transmission has sets of spinning gear wheels or cogs. Some are on the layshaft, an extra shaft between the drive shaft from the engine and the shaft to the road wheels. The gear change mechanism slides the gear collar along a shaft so it rotates to make the cogs fit together in different combinations. This makes the road wheels turn faster or slower for the same engine speed.

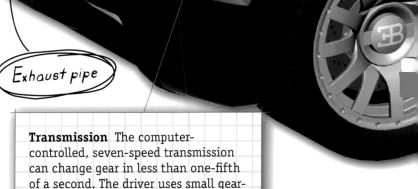

The collar (purple) slides along ridges on the drive shaft. Selecting high or low gears locks the teeth on the side of the collar onto either gear wheel

Driver's gear lever

The selector slides the collar between gears

Low gear selected

High gear

Drive shaft

Drive to wheels

Drive from engine

The gear wheels are driven from the layshaft and spin free on the drive shaft until locked on by the collar

Layshaft

Exhaust pipe

Transmission The computer-controlled, seven-speed transmission can change gear in less than one-fifth of a second. The driver uses small gear-shift paddles next to the steering wheel.

To learn more about transmissions, visit
www.factsforprojects.com and click the web link.

The Bugatti Veyron has ten radiators, including three for the engine and two for the air conditioning!

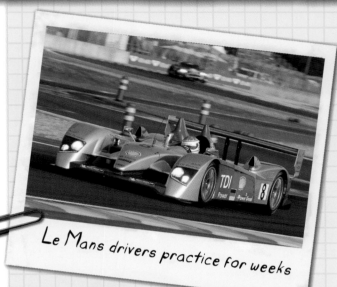

Le Mans drivers practice for weeks

LE MANS 24 HOUR

One of the world's most famous races, the Le Mans race for sports cars, lasts 24 hours nonstop. Only one car is allowed per team, but three drivers can take turns, although no one driver can stay at the wheel for more than four hours at a time. The cars come into the pits more than 30 times and cover more than 3,100 miles (4,989 km) at average speeds greater than 125 miles an hour (201 km/h)!

Brakes The brake discs are carbon composite. The pistons that push on them are titanium metal so they are less affected by great heat.

BUGATTI VEYRON

Bugatti has made many amazing cars over the years. The massive Royale of the 1920s had a 12-liter engine and a hood longer than many modern small cars.

Propeller shaft In a mid- or rear-engined four-wheel-drive sports car, the prop shaft carries the engine's turning power to the front wheels.

Half-shaft Each road wheel has its own short axle, or half-shaft. The Veyron is a four-wheel drive vehicle.

Wide low-profile tires

Alloy wheels

F1 RACING CAR

Formula One cars are not the largest, or the fastest, or the most powerful. But for all-round performance on a twisty track that requires speeding up and then braking hard to scream around corners, they cannot be beaten. An F1 car is built according to more than 1,000 rules and regulations, from engine size to overall weight, the electronic sensors it must have, and using the same transmission for four races in a row.

Eureka!

After many kinds of races with different cars and rules, the first F1 season was held in 1950. There are about 18 races around the world in a year. Each race is more than 190 miles (306 km) long but lasts less than two hours.

What next?

The Rocket Racing League plans to hold races for rocket-powered cars and aircraft. Each race could last between 60 and 90 minutes.

Telemetry Sensors for speed, brake temperature, and many other features send information by radio signals to the team members in the pits.

Suspension arm The suspension arms swing to allow the wheel to move up and down.

Mirrors

An F1's engine is part of the car's structure. It is bolted to the driver's cockpit at the front, the transmission, and rear suspension.

Front wing The front wing produces about one-third of the down force of the rear wing. It keeps the front tires pressed hard onto the track for accurate steering. The upright end plates direct air smoothly over the wheels.

Nose cone

ECU (electronic control unit) receives signals from sensors and adjusts length of each spray pulse

Fuel injector squirts fuel into air entering cylinder

Electricity supply

Air enters cylinder

Cylinder

Fuel pump

Fuel tank

Fuel pressure regulator allows unused fuel back to the tank

✳ How does FUEL INJECTION work?

A fuel injector squirts fuel, under pressure from a fuel pump, into air being drawn into the cylinder. An electronic control unit calculates how much fuel per squirt, depending on sensor information such as air pressure, engine speed, and how much oxygen is in the exhaust gases (which is linked to how much fuel is left unburned).

Tires There are tires for dry conditions, wet conditions, and intermediate (in between). Dry tires are slicks with no tread pattern.

Watch amazing videos of Formula One races by visiting www.factsforprojects.com and clicking the web link.

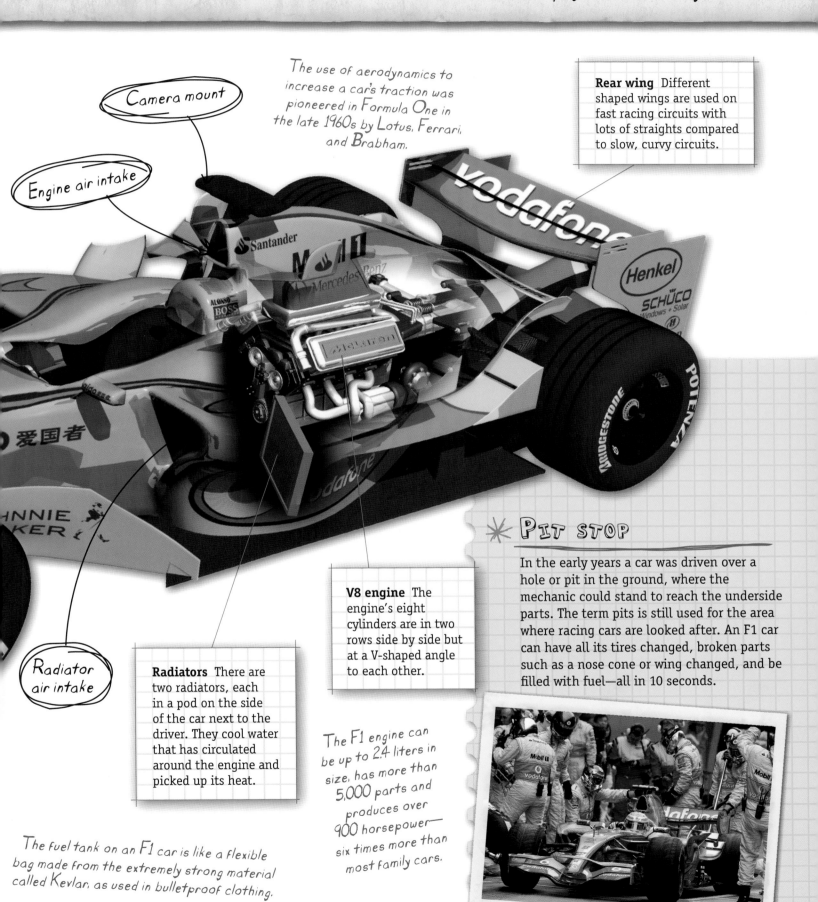

Camera mount

Engine air intake

The use of aerodynamics to increase a car's traction was pioneered in Formula One in the late 1960s by Lotus, Ferrari, and Brabham.

Rear wing Different shaped wings are used on fast racing circuits with lots of straights compared to slow, curvy circuits.

Radiator air intake

Radiators There are two radiators, each in a pod on the side of the car next to the driver. They cool water that has circulated around the engine and picked up its heat.

V8 engine The engine's eight cylinders are in two rows side by side but at a V-shaped angle to each other.

The F1 engine can be up to 2.4 liters in size, has more than 5,000 parts and produces over 900 horsepower—six times more than most family cars.

The fuel tank on an F1 car is like a flexible bag made from the extremely strong material called Kevlar, as used in bulletproof clothing.

✳ PIT STOP

In the early years a car was driven over a hole or pit in the ground, where the mechanic could stand to reach the underside parts. The term pits is still used for the area where racing cars are looked after. An F1 car can have all its tires changed, broken parts such as a nose cone or wing changed, and be filled with fuel—all in 10 seconds.

Pit crew refuel Lewis Hamilton's car

DRAGSTER

Dragster racing is the world's loudest, fastest form of motor racing—yet each race has only two competitors, no corners, and lasts just a few seconds. The idea is to accelerate (pick up speed) as quickly as possible from a standing start and be first across the finish line either one-quarter of a mile or one-eighth of a mile away.

Eureka!

In 1951, Wally Parks had the idea of making unofficial and dangerous dragster-type street racing into an official sport. He founded the U.S. National Hot Rod Association, which has run the sport since.

What next?

Some drivers have experimented with jump ramps halfway along the drag strip (track) so that the race is half on the ground and half flying through the air!

Dragsters have just one gear. There's no time to change to a second one.

Streamlined body The lightweight body is long, slim, and tapering so that it slices through the air like an arrow.

Cockpit

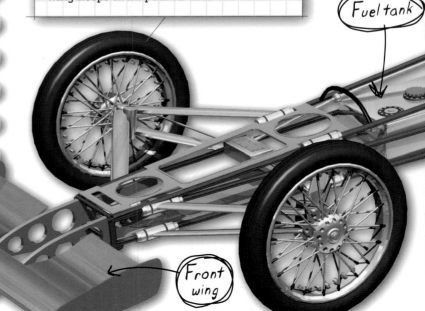

Front wheels The tiny front wheels mean less weight and air resistance. The front wing keeps them pushed onto the track.

Fuel tank

Front wing

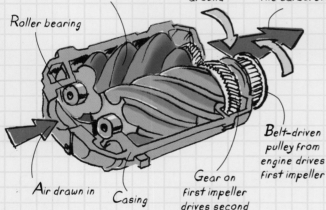

Twin meshing impellers draw in and compress the air

Impellers spin around

Air blasts into the carburetor

Roller bearing

Air drawn in

Casing

Gear on first impeller drives second impeller

Belt-driven pulley from engine drives first impeller

✳ How does a SUPERCHARGER work?

A supercharger forces air into the engine at high pressure so it carries extra fuel for greater speed and power. It consists of two screw-like devices called impellers that draw in air through an inlet and blast it from the outlet into the carburetor. The impellers are driven by a belt or chain from the engine. Turbochargers are similar but use a fanlike turbine rather than a direct mechanical drive (see page 28).

See drag racing in action by visiting www.factsforprojects.com and clicking the web link.

The huge rear tires on top-fuel dragsters wear out after about five races—less than 1.25 mi (2 km). Ordinary car tires usually last 18,500 mi (29,773 km) or more.

Smoke and flames at the burnout

✳ WHACKY RACERS!

There are dozens of types or classes of dragster racing, depending on engine size and type of fuel. Before the starts, the enormous rear tires are spun against the ground in a burnout while the dragster stays still to make them hot and sticky so they grip better.

Roll cage Drivers sit within a tubular metal frame or cage that gives protection if the dragster rolls over.

The fastest dragsters cross the finish line at more than 330 mph (531 km/h)—almost five times the U.S. freeway speed limit.

Rear wing

Air intake manifold

Instrument panel

Supercharger

Short stub exhausts

Tubular alloy chassis The light, but stiff, chassis (the main framework) is made of various alloys or mixtures of metals.

Top-fuel dragsters, the fastest type, use about 5 gal (18.9 l) of fuel during the race, which is about 800 times more than a family car would use.

Supercharged V8 engine The largest engine in most races is 8.2 liters. Its supercharger means it can produce more than 5,000 horsepower.

Rear wheels The huge rear wheels have soft, slick (treadless) tires. The driver and engine are both near the back so that their weight helps the tires to press down and grip.

4WD OFF-ROADER

A four-wheeled vehicle with four-wheel drive (4WD or 4x4) means that all four wheels are turned by engine power. A 4x2 vehicle has four wheels, but only two are engine powered. A 6x4 is a six-wheeled vehicle with engine power to four wheels. Four-wheel drive is best for ATVs (All-Terrain Vehicles) that can go off road and across almost any kind of terrain or ground, from soft sand to squashy mud to steep rocky slopes.

Eureka!

Car designers who built the first 4WD vehicles in the 1900s included Ferdinand Porsche, founder of the famous Porsche sports car and racing car organization. His first 4WD had an electric motor for each wheel. On some types of 4WDs, all four wheels steer, not just two.

What next?

Military off-road vehicles have tested 4WDs with an extra two wheels that swing down to give added grip in mud.

EXPLODED VIEW OF HUMVEE

The Humvee's engine air intake, exhaust, electrical wires, and similar parts are designed so that the vehicle can drive through water deeper than 3 ft (1 m).

Snorkel Air for the engine is drawn in through a tall pipe to avoid taking in water while crossing streams.

Camouflage Military vehicles, such as the Humvee, are painted so they blend in with their surroundings. This is called camouflage.

The military 4WD called the Humvee is used by U.S. and other military forces around the world.

Lights

Tow points

Radiator The radiator is well protected against damage from rocks by a strong metal plate underneath it.

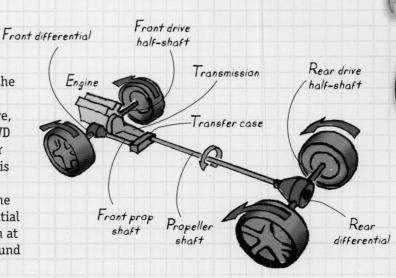

✳ How does 4WD work?

In some cars, a front engine drives the two rear wheels by a long propeller shaft. Most cars are front-wheel drive, with half-shafts at the front. The 4WD system combines the two, so all four wheels turn under engine power. This gives much better grip on rough or slippery surfaces, especially when the tires have a deep tread. The differential makes the two wheels it powers turn at different speeds, so when going around a corner, the outer one turns faster.

Front differential

Front drive half-shaft

Transmission

Rear drive half-shaft

Engine

Transfer case

Front prop shaft

Propeller shaft

Rear differential

Humvees are so tough they can be dropped by parachute from cargo planes.

Discover everything there is to know about the Humvee by visiting www.factsforprojects.com and clicking the web link.

Machine gun The rotating roof turret, or cupola, with its armored guard, can be fitted with a powerful machine gun.

Armor plating Thick, strong, light metal plates cover most of the bodywork to protect against bullets, land mines, and other dangers.

The Humvee is named from its initials, HMMWV, meaning High Mobility Multi-purpose Wheeled Vehicle.

High exhaust

The Hummer can tackle any kind of difficult terrain

✳ MUSCLE-MACHINE!

The Humvee 4WD is available in a civilian (nonmilitary) version—the Hummer 1. There are many other types of 4WDs used by farmers, foresters, ranchers, explorers, countryside workers, and of course drivers keen on off-roading. Actor and politician Arnold Schwarzenegger has a "green" Hummer that's been converted to run on nonpolluting hydrogen fuel.

Engine Military 4WDs such as the Humvee have a 6.5-liter diesel engine with fuel injection.

Chassis The main frame has two long girder-like rails and several cross members.

RALLY CAR

Rallies are tough races that can take place almost anywhere, from public roads (closed to everyday traffic for the event) to dirt tracks, forest trails, ice and snow, deserts, and real racing circuits. A rally car is a special version of a normal production car that has a tuned-up engine and stronger mechanical parts.

Eureka!

Before satnav (satellite navigation), which uses GPS (Global Positioning System of satellites), rally drivers sometimes got lost and ended up dozens of miles from the finish line.

The Dakar Rally is the longest, hardest race. It runs more than 6,000 mi (9,656 km) from European cities to Dakar, Senegal, in West Africa. Part of the race is across the Sahara Desert.

☀ ON THE SPEEDWAY

Stock cars are ordinary production cars with certain changes and modifications, as allowed by the rules, to compete on proper racing circuits. They roar around giant oval tracks called speedways as part of the NASCAR season—the National Association for Stock Car Auto Racing. NASCAR drivers can reach speeds of more than 185 miles an hour (298 km/h).

Stock cars on the NASCAR "bowl"

Tough suspension The springs, shock absorber, and other suspension parts take a huge hammering at rallies.

Rear differential The differential can be locked to make both rear wheels turn at the same speed for getting out of holes and ditches.

Lowered body

Rally cars are 4WD and based on 2-liter turbocharged engines.

Brakes Heavy duty brakes mean that rally drivers can brake at the last split second as they enter corners to clock the fastest time.

What next?

Inventors have built amphibious cars that have wheels for normal road conditions, plus floats with propellers to travel through water like a boat.

Learn about different types of steering systems by visiting www.factsforprojects.com and clicking the web link.

Roll cage A framework of strong tubes inside the passenger compartment stops the sides or roof caving in if there's a crash.

The World Rally Championship consists of about 15 races all around the world.

Internal padding All hard objects near the driver and codriver are padded to avoid injury when bouncing along rough roads.

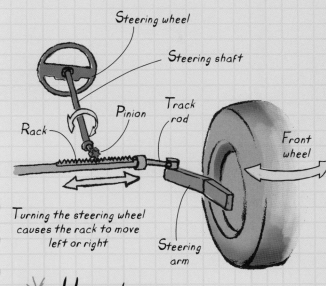

Steering wheel

Steering shaft

Pinion

Track rod

Rack

Front wheel

Steering arm

Turning the steering wheel causes the rack to move left or right

✳ How does STEERING work?

A car's steering wheel is fixed to a long shaft called a steering column with a small gear called a pinion at its base. As the pinion turns it makes a rack—a long bar with teeth—slide left or right. Each end of the rack is linked to a smaller bar known as a track rod, which is attached by a ball joint to another bar, the steering arm, and this is attached to each front wheel hub. As the rack slides left or right, it moves the track rod and steering arm. The steering arm works as a lever to make the front wheels angle left or right.

Steering rack and pinion

Tuned transverse engine The engine is carefully adjusted, or tuned, so that it runs with the greatest power yet does not use too much fuel. This saves fuel weight and also reduces the number of refueling stops.

Alternator (generator)

Spotlights

PICKUP TRUCK

If you want to transport a heavy load, a pickup truck is the ideal vehicle. These small but tough trucks have an open flat area called a load bed for their cargo. Some are two- or three-seaters with one row of seats. Others have a second row behind the driver. The strengthened, stiffened rear suspension means the ride is not as comfortable as an ordinary car.

Pickup racing is a fast and furious motor sport in which the modified trucks can speed along at more than 120 mph (193 km/h).

Eureka!

In the early years of motoring, people cut the rear body off a car and added a wooden platform to make a pickup truck. The first mass-produced versions, based on the Ford Model T, were sold in 1925.

What next?

Some pickups have a topper that fits in the bed of the truck. The topper can include sleeping and cooking areas.

One traveling circus in the USA had an elephant specially trained to ride on the back of a pickup truck.

Engine Most pickups have diesel engines. These are heavy and noisy but powerful and easy to adjust and maintain.

Tinted glass

✳ How does a TURBODIESEL work?

A diesel engine is similar to a gasoline engine (see page 16) but it lacks spark plugs. The air-fuel mixture explodes in the cylinder because it gets hot from being compressed. The turbocharger, or turbo, is similar to a supercharger (see page 22), but the impeller that forces extra air into the engine is worked by a fanlike turbine spun around by exhaust gases.

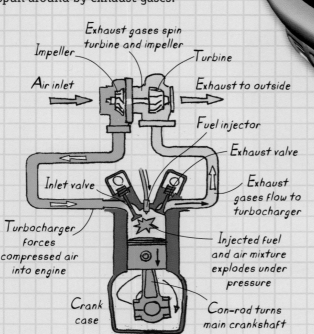

Exhaust gases spin turbine and impeller

Impeller

Turbine

Air inlet

Exhaust to outside

Fuel injector

Exhaust valve

Inlet valve

Exhaust gases flow to turbocharger

Turbocharger forces compressed air into engine

Injected fuel and air mixture explodes under pressure

Crank case

Con-rod turns main crankshaft

Foglights

Turbo

Tires Pickups have thick, wide, strong tires to spread the truck's weight, and tread to grip soft ground.

In Australia, pickup trucks are often called "utes" (utility vehicles).

To learn about modified pickup trucks with monster wheels,
visit www.factsforprojects.com and click the web link.

Crew cab The rear seats give extra room for the crew—people who help to load and unload the truck. However they reduce the area of the load bed behind them.

Load bed The cargo platform is usually made of metal with ridges for strength. Hooks for ropes and straps stop the load from sliding about.

In South Africa, pickups are commonly known as "bakkies" from their similarity to a metal baking tin for loaves of bread.

Lights Extra-bright front and rear lights help the driver to see when collecting loads in remote places such as farms.

Exhaust

Prop shaft

Muffler

Running board A flat strip along the lower side helps stepping up into the truck, which is higher than an ordinary car.

✳ LOADMASTERS

Pickups are very adaptable working vehicles because the load bed can take objects of different sizes and shapes. In the rain, a waterproof sheet called a tarp is tied over the load to keep it dry. A "half-tonner" pickup can safely transport a load of 1,000 pounds (453 kg). Most of the larger versions have a one ton (0.9 metric tons) carrying capacity.

Pickups can carry almost any cargo

CITY BUS

There are many kinds of passenger-carrying buses and coaches for different services. Some carry fewer people long distances in comfort with soft seats and lots of legroom. Others pack in as many people as possible, often standing up, for short trips around towns and cities. In some places, electric buses are replacing diesel-engined ones to keep city streets quieter and the air cleaner.

Eureka!

The earliest buses in the 1700s were horse-drawn wagons with two benches along the middle. The passengers sat back-to-back, facing sideways. There were no sides or roof to keep out the wind and rain.

What next?

The latest long-distance buses have screens for computers or movies and earphones for music, like a passenger airplane.

Automatic doors
The driver works buttons that make the passenger door swing open using an electric mechanism.

One of the world's biggest buses is the Superliner from Shanghai, China. It is 80 ft (24.4 m) long and can carry up to 300 passengers. It bends to go around corners.

Rack and pinion steering

✳ BENDY BUSES

Many old cities have narrow streets and sharp corners unsuitable for long buses. The articulated (jointed) or "bendy" bus has a link in the middle so it can turn corners more tightly than a rigid one-piece vehicle. Some bendy buses have two links joining three sections. The driver keeps watch on the rear end using closed-circuit television cameras (CCTV) and a screen.

Driver-only Most modern buses are driver-only. The driver collects the money and gives out tickets. Some buses have a conductor, who collects the fare, and a driver.

An articulated bus in London

>>>CARS, TRUCKS, AND BIKES<<<

Read facts and view pictures of many different kinds of buses by visiting www.factsforprojects.com and clicking the web link.

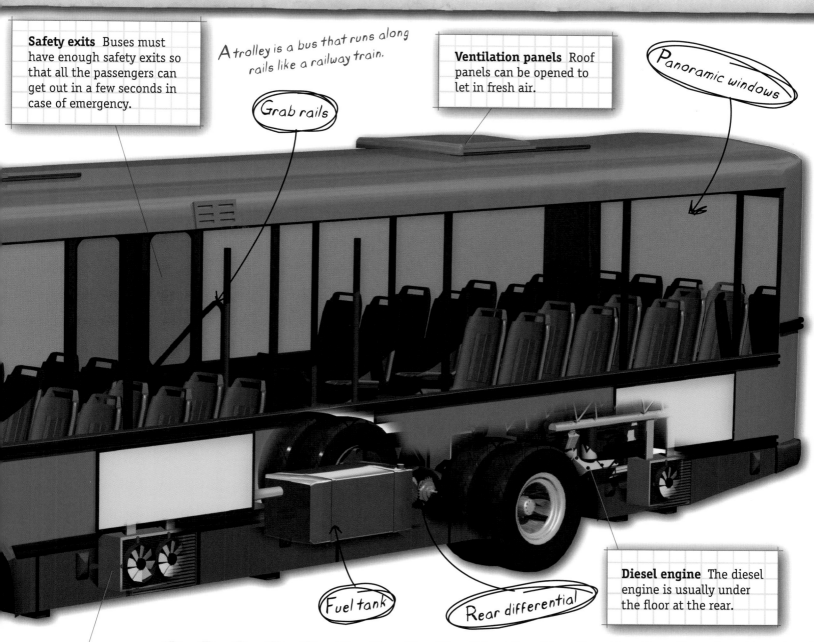

Safety exits Buses must have enough safety exits so that all the passengers can get out in a few seconds in case of emergency.

A trolley is a bus that runs along rails like a railway train.

Grab rails

Ventilation panels Roof panels can be opened to let in fresh air.

Panoramic windows

Fuel tank

Rear differential

Diesel engine The diesel engine is usually under the floor at the rear.

Air conditioning The air inside the bus is heated or cooled depending on the outside temperature.

A trolley bus is an electric bus that gets its electricity from long "arms" that touch overhead wires.

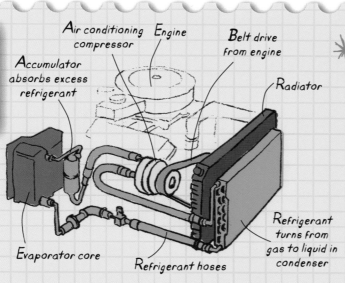

Air conditioning compressor

Engine

Belt drive from engine

Accumulator absorbs excess refrigerant

Radiator

Evaporator core

Refrigerant hoses

Refrigerant turns from gas to liquid in condenser

✳ How does AIR CONDITIONING work?

Air conditioning, or climate control, works in a similar way to a refrigerator. A compressor squeezes a gas, the refrigerant, around a circuit of pipes. The compressed gas condenses and becomes a liquid and in the process gets hot. In the second part of the circuit, the pressure is lower. The liquid refrigerant expands and evaporates (turns back into a gas) and in the process becomes much colder.

SEMITRUCK

A semitruck is articulated, or jointed. The joint is between the front part—the tractor unit with the engine and driver's cab—and the rear part, or trailer, which carries the load. The joint allows the truck to go around tighter corners than a one-piece vehicle. It also means different kinds of trailers can be joined, or hitched, to the tractor unit.

Eureka!

The first articulated trucks were built in the 1910s by Charles Martin. He hitched a tractor-like truck to a wagon usually pulled by horses. He also invented the fifth wheel coupling between the towing unit and trailer (below).

What next?

Most countries limit the size of trucks by their weight or length. However, new super-highways could handle trucks of 100 tons or more.

The ShockWave truck of Hawaii's Fire Department has two jet engines and can reach speeds of 300 mph. It's only used for shows, not to race to real fires.

Trailer

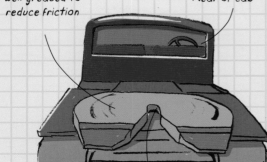

Fifth wheel is well greased to reduce friction

Rear of cab

Trailer's king pin engages in slot

☀ How does the FIFTH WHEEL work?

The fifth wheel coupling is the joint or link between the tractor unit and trailer. It consists of a king pin or coupling pin on the lower front of the trailer that slides up on and then slots into the U-shaped fifth wheel on the rear of the tractor unit. The trailer can swing from side to side behind the tractor unit and also move by a small amount up or down to cope with bumpy roads.

Trailer stands
When the trailer is unhitched, its front leans on these strong metal legs.

The Centipede truck is 180 ft (54.9 m) long and weighs 225 tons (204 metric tons)— the longest truck in regular work.

Find out everything there is to know about ice road truckers by visiting www.factsforprojects.com and clicking the web link.

A road train is one tractor unit pulling several trailers, like a railway locomotive pulls several cars. Some road trains are more than 3,000 ft (914.4 m) long and weigh more than 1,000 tons (907 metric tons).

Roof fairing Even on a 40-ton (36.2 metric tons) truck, smooth streamlining helps to lower air resistance. This increases speed and reduces fuel use.

Cab controls A big truck has ten gears for all conditions—from cruising the open road with no load to climbing a steep hill with 30-plus tons (27 metric tons) of cargo on the back.

Towing unit

Sun visor

Fifth wheel

King pin

Engine Truck turbodiesel engines are 11, 13, or 16 liters, and sometimes even more.

VOLVO

520

Fuel tank A family car's fuel tank holds about 20 gallons (75 l). Large semitrucks can carry from 130 to 500 gallons (492-1,893 l).

✳ ICE ROAD TRUCKERS

Sometimes it's quicker for a car or truck to get to a remote place across a lake—provided it's frozen. Ice truckers specialize in carrying loads across the far north in winter, to faraway places such as mining centers and logging camps. The truckers keep in radio contact with each other about snowdrifts, cracks, or melting ice.

A semitruck crosses a frozen lake

33

TOW TRUCK

Everyone on the road fears a sudden breakdown, but sometimes it can't be avoided. Soon after mass motoring began in the 1900s, specialized trucks rescued stranded drivers and towed their broken vehicles. It's important to clear the road and get the vehicle out of danger, then take it for repair. This may have to be done at night, in heavy rain or in thick fog or snow. Tow trucks are strong, tough, and able to cope with all conditions.

Eureka!

Mechanic Ernest Holmes built the first tow truck in 1915. He fixed three metal poles, a chain, and a pulley to a 1913 Cadillac in Chattanooga, Tennessee, and began the tow truck business.

What next?

Motoring experts are working on intelligent vehicle electronics to sense which part has broken and radio the tow truck to bring it as soon as possible.

Powerful engine The turbodiesel engine must be powerful enough to move two vehicles, perhaps across soft ground, if the disabled one has veered off the road.

The world's biggest tow trucks are converted Caterpillar 793s used in mines to recover giant haulage trucks weighing over 440 tons (399 metric tons).

Visor

Air filter

✳ How does a WINCH work?

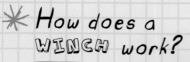

A winch has a strong metal cable or wire that winds slowly onto a drum. Some tow truck winches are electric with a powerful motor driven by the truck's battery, or the truck's engine. The turning speed is greatly reduced by gear cogs. As the turning speed goes down, the turning force or torque increases. The cable winds very slowly but with huge force to drag or lift the broken-down vehicle.

Counterweight A very heavy vehicle at the back of the truck might make the front end lift up, so the counterweight keeps the front end down.

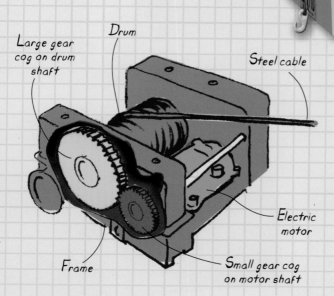

Large gear cog on drum shaft

Drum

Steel cable

Electric motor

Frame

Small gear cog on motor shaft

The International Towing and Recovery Hall of Fame and Museum is in Chattanooga, Tennessee, near where Holmes started the first tow truck business.

The AAA (American Automobile Association) started in 1902 in Chicago. It was a response to the bad roads that caused cars to break down regularly.

Read about the history of vehicle recovery by visiting www.factsforprojects.com and clicking the web link.

Flashing lights

Winch The steel cable from the hoist hook winds slowly onto the winch drum.

Boom The boom (lift arm or crane) is worked by two powerful hydraulic jacks. Its L-shaped end can be slotted underneath the front of a vehicle to raise it from beneath, instead of using the hook.

Hook

Ramps The rear ramps can be lowered to support the broken vehicle's front wheels.

Eight rear wheels spread the load

Step

Tools A large tool compartment contains spanners, crowbars, screwdrivers, cutters, and other essential equipment.

A flatbed (rollback or slide) truck recovers a disabled car

✷ SLIDING UP AND ON

There are several kinds of tow trucks. Some have cranelike booms to lift vehicles straight up out of ditches or rivers. Some have an arm and winch to drag the broken vehicle to safety. Others have a flat rear platform or flatbed that slides backward and down onto the road, so the vehicle can be winched up onto it. Then the platform slides back up.

FIRE TRUCK

Firefighting trucks or fire engines may be first to arrive at big incidents—and not just fires, but floods, road accidents, people trapped in holes or in high places, and even kittens in trees. The fire truck sprays water or special types of chemical foam depending on the type of fire and what is burning, such as wood, plastic, or fuels.

Eureka!

The first firefighting wagons were hauled by people, horses, or both. Self-propelled trucks were powered by steam engines in New York from 1841 and in London and other British cities by the 1850s.

What next?

Fire crews carry out regular tests on new siren sounds—although they warn people they are nearby—one sounds like a whining dog!

Flashing lights

Siren The siren is worked by an electric motor and fan that pumps air past a specially shaped hole into a tube, similar to blowing a trumpet.

✳ EXTENDING LADDERS

Multisection ladders that extend are mounted on a turntable on the fire vehicle. They can reach up to 100 feet— the 10th floor of a high-rise building. The crew member at the upper end of the ladder is in radio contact with the ladder operator below so that the ladder can be put into the correct position. Spraying water or foam onto a fire from above is far more effective than from the side.

Fire and ambulance crews respond to emergency calls.

Engine The diesel engine is started regularly to make sure it will work when needed.

The extending ladder reaches above the fire

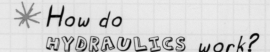

Take an on-line tour of a fire engine by visiting
www.factsforprojects.com and clicking the web link.

Hoses One set of hoses connects to a nearby water supply to draw water in. Another set carries the water away to spray on the fire. The hoses wind onto reels turned by electric motors.

Mains supply pipes

Screw fittings link pipes

✳ How do HYDRAULICS work?

Hydraulic machinery, such as cutters or extending ladders, works using high-pressure liquid (water or oil). Like a lever, the liquid changes a small force moving a long distance into a big force moving a short distance. The small force presses on a narrow piston to create pressure throughout the fluid. This pushes a wider piston with greater force because of its larger surface area.

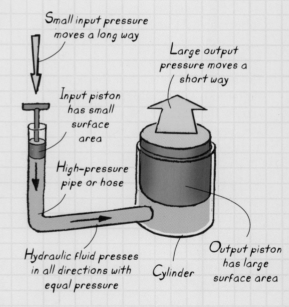

Small input pressure moves a long way

Large output pressure moves a short way

Input piston has small surface area

High-pressure pipe or hose

Hydraulic fluid presses in all directions with equal pressure

Cylinder

Output piston has large surface area

Wheel hub

New York City's fire department is the largest combined fire and emergency medical unit in the world.

Control panel The switches and other controls are for the main pumps that force water or foam along the hose, and display its pressure and flow rate.

Some fire trucks have pumps so powerful they can spray a distance of more than 230 ft (70 m).

Tools A fire truck carries a host of useful tools including powerful hydraulic cutters and spreaders operated by high-pressure hoses from the diesel engine.

EXPLODED VIEW

GLOSSARY

Alloy

A combination of metals or metals and other substances for special purposes such as great strength, extreme lightness, resistance to high temperatures, or all of these.

Articulated

A joint or flexible part rather than being rigid all the way along.

Bearing

A part designed for efficient movement to reduce friction and wear, for example, between a spinning shaft or axle and its frame.

Chassis

The main structural framework or skeleton of a car, which gives it strength and to which other parts are fixed, such as the engine and seats.

Cogs

The name for the teeth of a gear wheel or for whole gearwheels (which are sometimes called cogwheels).

Con-rod

Connecting rod, an engine part that links the piston to the main crankshaft.

Crank

A bent part of a spinning shaft or axle, or an armlike part sticking out from it, like the crank of a bicycle with the pedal fitted to it.

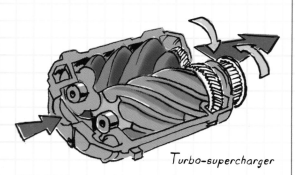
Turbo-supercharger

Disc brake

Crankshaft

The main turning shaft in an engine that is made to rotate by the up-and-down movements of the pistons.

Cylinder

The chamber inside an engine, in which a well-fitting piston moves.

Damper

A part that reduces or dampens movements, usually sudden jolts or vibrations. On vehicle suspensions, also called the shock absorber.

Derailleur

The gear-change mechanism on many bicycles that moves the chain sideways so it goes from one sprocket to another. It also takes up slack along the length of the chain to keep it taut when it moves between different-sized sprockets.

Diesel engine

An internal combustion engine (one that burns or combusts fuel inside a chamber, the cylinder) that uses diesel fuel and causes this to explode by pressure alone rather than by a spark plug.

Differential

A part that makes road wheels turn at different speeds as a vehicle goes around a bend. The wheel on the outside of the curve must spin slightly faster because it has farther to go than the inner wheel and would otherwise skid.

Disc brake

A system where two stationary pads or pistons press onto either side of a rotating flat disc attached to the wheel to slow it down.

Friction

When two objects rub or scrape together causing wear and losing movement energy by turning it into sound and heat.

Gasoline engine

An internal combustion engine (one that burns or combusts fuel inside a chamber, the cylinder) that uses gasoline and causes this to explode using a spark plug.

Gears

Toothed wheels or sprockets that fit or mesh together so that one turns the other. If they are connected by a chain or belt with holes where the teeth fit, they are generally called sprockets. Gears are used to change turning speed and force, for example, between an engine and the wheels of a car or to change the direction of rotation.

Gyroscope

A device that maintains its position and resists being moved or tilted because of its movement energy, usually consisting of a fast-spinning ball or wheel.

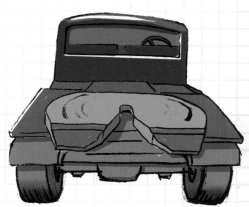
Fifth wheel

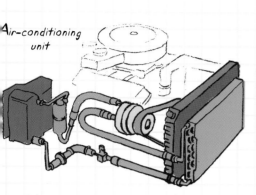

Air-conditioning unit

Hybrid car

A car with more than one method of moving, such as a gasoline engine and an electric motor.

Hydraulic

Working by high-pressure liquid such as oil or water.

Impeller

A spinning part similar to a fan or turbine with angled blades, which forces a gas or liquid into an area at high pressure.

Muffler

Part of an exhaust system that makes the noise of the exploding gases from the engine quieter.

Piston

A wide, rod-shaped part, similar in shape to a soda can, that moves along or up and down inside a close-fitting chamber, the cylinder.

Rack

In gear systems, a long strip with teeth or cogs, as used in rack-and-pinion steering.

Radiator

In cars and similar vehicles, a part designed to give off heat, for example, from an engine. It has a large surface area. Hot water from the engine circulates through it to become cooler before flowing back to the engine.

Satnav

Satellite navigation using radio signals from the GPS (Global Positioning System) satellites in space to determine locations. Many cars have in-built satnav systems or small, portable gadgets.

Slicks

Tires that have little or no pattern of grooves, or tread, usually used on dry road surfaces. Slicks are used by racing cars from drag racers to Formula 1.

Sprocket

A wheel with teeth around the edge, often called a gear wheel. Unlike gears, sprockets do not fit together or mesh directly, but have a chain or belt between them.

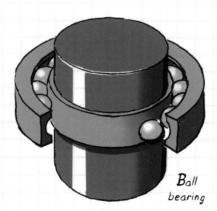

Ball bearing

Suspension

Parts that allow the road wheels of a vehicle to move up and down separately from the driver and passengers to smooth out bumps and dips in the road. Also, any similar system that gives a softer, more comfortable ride.

Throttle

A control that allows more fuel and air into the engine for greater speed, sometimes called an accelerator.

Towing unit

In an articulated truck, the part with the engine or cab that does the pulling, also called the prime mover.

Transmission

The parts that transmit the turning force from the engine (crankshaft) to the wheel axles, including the gears, transmission, and propeller shaft.

Suspension strut

Turbine

A set of angled fanlike blades on a spinning shaft that is used in many areas of engineering, from pumps and cars to jet engines.

Turbo

An engine, pump, or similar device that works using a turbine.

Valve

A part that controls the flow or passage of a substance, similar to a faucet or the movement of fuel and air mixture into an engine.

Winch

A winding mechanism that slowly turns or reels in a rope or cable, with great force.

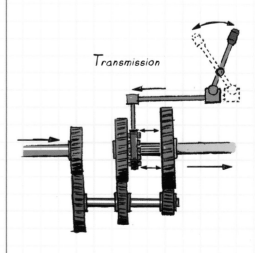

Transmission

INDEX

AAA (American Automobile Association) 34
air conditioning 19, 31, 39
alloys 11, 15, 17, 19, 23, 38
alloy wheels 15, 17, 19
alternators 27
amphibious cars 27
armor plating 25
articulated trucks **32–33**, 38, 39
Aston Martin sedan 17
ATVs (All-Terrain Vehicles) 24
Austin Mini 6, 17
automatic doors 30
axles 17, 19, 38, 39

ball bearings 10, 39
batteries 17, 34
bearings 38
bendy buses 30
Benz, Karl 16
bicycles 6, 7, 8, 9, 10
boom 35
Brabham 21
brake discs 13, 14, 19
brakes 7, 9, 13, 14, 19, 26, 38
Bugatti Veyron 18, 19
burnout 23
buses **30–31**

camouflage 24
carbon fiber 11
carburetors 22
Caterpillar 793 34
chain and sprocket gears 8
chains 6, 8, 38, 39
chassis 23, 25, 38
city buses **30–31**
clutch levers 14
cockpits 22
cogs 18, 38
compressors 31
computers 16
con-rods 16, 28, 38
control panels 37
counterweights 34
cranks 38
crankshafts 16, 28, 38, 39
cupolas 25
cylinders 12, 16, 18, 20, 21, 28, 37, 38, 39

Daimler, Gottlieb 12
Dakar Rally 26
derailleurs 8, 38
differentials 24, 26, 38
diesel engines 25, 28, 30, 31, 36, 38
disc brakes 9, 14, 38
dragsters 7, **22–23**
drive belts 12
drive shafts 18, 24
Dunlop, John Boyd 10

electric motors 6, 9, 13, 14, 17, 24, 30, 34, 36, 37, 39
engines 6, 7, 12, 13, 14, 15, 16, 18, 19, 20, 21, 22, 23, 24, 25, 26, 28, 31, 32, 33, 34, 36, 38, 39
exhaust 12, 15, 16, 18, 23, 24, 25, 28, 29, 39

F1 (Formula One) racing cars 7, **20–21**
Ferrari 21
fifth wheel 32, 33, 38
fire engines **36–37**
flatbeds 35
Ford Model T 17, 28
Ford Motor Company 6
forks 13
four-wheel drive off-roaders (4WD) **24–25**, 26
frames 6, 9, 11, 15, 38
front wheel drive 17, 19
fuel 16, 20, 21, 22, 23, 25, 27, 33, 38, 39
fuel injection 20, 25, 28
fuel tanks 13, 21, 31, 33

gasoline engines 12, 13, 16, 17, 28, 38, 39
gear cables 8, 9
gear changers 8, 9, 11, 18, 38
gear cogs 34
gears 6, 7, 11, 12, 15, 17, 18, 22, 27, 33, 38, 39
GPS (satellite navigation) 16, 26, 39
green cars 17
gyrobikes 9
gyroscopes 9, 38

half-shafts 19, 24
Hamilton, Lewis 21
Harley Davidson touring motorcycle **12–13**
Harley, William 13
Hildebrand & Wolfmüller 12
Holmes, Ernest 34
hoses 37
Howard, John 9
Hummer 25
Humvee 4WD 24, 25
hybrid cars 17, 39
hydraulics 35, 37, 39
hydrogen fuel 25

ice road truckers 33
impellers 22, 28, 39

king pins 32, 33

ladder, extending 36, 37
layshafts 18
Le Mans 24 hr race 19

load beds 28, 29
Lotus 21

Martin, Charles 32
maxi-scooters 13
motorcycles 6, **12–13**, **14–15**
mountain bikes **8–9**, 10
mudguards 13
muffler 29, 39
Murphy, Charles Minthorn 11

NASCAR (National Association for Stock Car Auto Racing) 26
National Hot Rod Association 22
nose cone 20

off-roaders, 4WD **24–25**
Olympics 8, 9

Parks, Wally 22
pedals 6, 8, 10, 11, 14, 15, 38
penny-farthing bicycles 6
pickup trucks 7, **28–29**
pinion 7, 27, 30, 39
pistons 14, 16, 19, 37, 38, 39
pit stop 21
Porsche, Ferdinand 24
propeller shafts 17, 19, 24, 29, 39

quad bikes 15

racing handlebars 10
racks 7, 27, 30, 39
radiators 14, 19, 21, 24, 31, 39
ramps 35
rear differentials 26, 31
rear transmission 17
rear wheel drive 17
rally cars 26-27
Reitwagen 12
road race bicycles **10–11**
Rocket Racing League 20
roll cages 23, 27
running board 29

saddles 11
safety exits 31
satnav (satellite navigation) 16, 26, 39
sedan cars **16–17**
shock absorbers 26, 38
ShockWave truck 32
sirens 36
slicks 15, 20, 23, 39
spark plugs 16, 28
speedway 26
spoilers 18

spokes 11, 15
spring suspension 8, 12
sprockets 6, 8, 11, 38, 39
steam engines 6, 36
steering 7, 14, 15, 16, 18, 20, 27
stock cars 26
Stoeckl, Markus 9
superbikes **14–15**
superchargers 22, 23, 28
Superliner buses 30
super sports cars **18–19**
suspension 7, 15, 16, 20, 26, 28, 38, 39
swingarms 15

telemetry 20
throttles 13, 39
throttle valves 20
tires 9, 10, 15, 18, 19, 20, 21, 23, 28, 39
tools 35, 37
touring motorcycles **12–13**
tow truck 7, **34-35**
towing unit 32, 33, 39
Toyota Corolla 17
track racing bikes 11
track rods 27
trailers 32, 33
transmission 12, 17, 18, 20, 24, 39
transverse engines 17, 27
trucks **28–29**, **32–35**
turbines 22, 39
turbochargers (turbos) 22, 26, 28, 38, 39
turbodiesels 28, 33, 34

V8 engines 21, 23
ventilation panels 31
Veyron, Pierre 18
VW Beetle 6

W16 engines 18
wheel rims 11
wheels 6, 7, 8, 11, 14, 15, 16, 17, 18, 20, 22, 23, 24, 26, 35, 37, 38, 39
winches 34, 35, 39
windshields 14
wings 18, 20, 21, 22, 23
World Rally Championship 27